Mister Moon

The excitement of searching for the moon each night & Discovering its many amazing phases, shapes, and colors

Written by Ivy E Jeltema

Illustrated by Rosalia Febyola

ISBN: 978-1-7378257-0-8

MISTER MOON
Dedicated to our little Boo Boo and the
memories of always searching for Mister Moon –I.J.

TEA

Mister Moon Mister Moon
Are you there tonight?
Hanging in the sky, round and bright?

I've searched and looked, still no sign
But wait, you appear with the faintest line
You are the new moon as you are dark
Your outline shines, crisp and stark

Mister Moon Mister Moon
Where are you tonight?
What will you bring for our evening light?

Easy to find with your brighter glow
Thank you for giving me more to show
You are the crescent moon now in phase
Displaying more of you for me to gaze

Mister Moon Mister Moon
Where are you tonight?
Should I look in the distance, left or right?

I discover you beyond the trees
Floating and rising in the breeze
You are the half moon, shiny and clear
Revealing full shape in shadow, craters and sphere

Mister Moon Mister Moon
What will I see of you tonight?
Will you brighten a path as my bedtime lite?

Exposing your shape from afar
The waxing gibbous is what you are
Extraordinary and unique, you're surprising today
Your surface beams, craters dark grey

Mister Moon Mister Moon
Are you there tonight?
Hanging in the sky, round and bright?

Your grand, warm face has finally arrived
Glimmering and illuminating the night's sky
Wondrous excitement, you are the full moon
Anticipating your return to the new phase soon

Goodnight Mister Moon!
See you tomorrow!

Mister Moon Mister Moon
Why are you different all of the time?
Closer, then further, then you climb
Red, orange, pink, yellow, silver, blue
Your lunar colors are constant and true

MOON PHASES
New Moon
Waxing Crescent
Half First Quarter
Waxing Gibbous

Full Moon
Wanning Gibbous
Half Last Quarter
Wanning Cescent

Every evening young children from around the world love seeing the moon in the night's sky. Their curiosity is powered by the moon's ever-changing appearance, varied illumination, and location as it gradually passes through all the lunar phases every 29 days. This book captures one child's journey & excitement as she searches and discovers a different "Mister Moon" each night. Color, shape, size; what will he reveal tonight?

$10.95
ISBN 978-1-7378257-0-8
51095>
9 781737 825708

Made in United States
Troutdale, OR
11/03/2026

56844428R00017